AVERY WRIGHT

From Predator to Phantom

A History of Drones". Write the introduction

First edition

This book was professionally typeset on Reedsy.
Find out more at reedsy.com

Contents

Template

Known specs

- **Unit Cost:**
- **Thrust:**
- **Wingspan:**
- **Length:**
- **Height:**
- **Maximum takeoff weight:**
- **Maximum speed:**
- **Range:**
- **Fuel Capacity:**
- **Ceiling:**
- **Endurance:**
- **Crew (remote):**
- **Payload capacity:**
- **Armament:**

History and usage

Additional information

1

Introduction

Drones, also known as unmanned aerial vehicles (UAVs), have come a long way since their first military use in the early 2000s. From the MQ-1 Predator to the DJI Phantom, drones have evolved in terms of technology and usage. They have been used for a variety of purposes, including military operations, scientific research, and civilian applications.

In this book, "From Predator to Phantom: A History of Drones," we will take a journey through the history of drones, from their early military use to their current civilian applications. We will focus on specific drones, including the MQ-9 Reaper, RQ-4 Global Hawk, MQ-1 Predator, RQ-21 Blackjack, IceBridge,

TDR-SS, DJI Phantom, DJI Mavic, Parrot Bebop, Yuneec Typhoon, and Autel Robotics EVO. Each chapter will provide known specs, history and usage of the drone, along with additional information.

The book will be divided into three sections: military drones, scientific drones, and civilian drones. By the end of this book, you will have a comprehensive understanding of the history, technology, and usage of drones, and how they have evolved over time.

Overview of the history of drones

The history of drones dates back to the early 20th century, where the first unmanned aerial vehicles (UAVs) were primarily used for reconnaissance and surveillance purposes. However, it wasn't until the early 2000s that drones began to be used for military operations. The MQ-1 Predator, developed by General Atomics, was the first drone to be used in combat by the United States Air Force. This marked the beginning of the widespread use of drones in military operations.

As technology advanced, drones began to be used for other purposes, such as scientific research and civilian applications. Scientific drones, such as IceBridge, have been used for mapping and studying the earth's polar ice caps. Civilian drones, such as the DJI Phantom, have been used for a variety of purposes, including photography, videography, and delivery services.

Explanation of the different types of drones

Military drones, also known as unmanned combat aerial vehicles (UCAVs), are used for reconnaissance, surveillance, and military operations. They are usually larger and more advanced than scientific and civilian drones. Examples of military drones include the MQ-9 Reaper, RQ-4 Global Hawk, Predator B, and MQ-1 Predator.

Scientific drones, also known as unmanned scientific aerial vehicles (USAVs), are used for scientific research and exploration. They are typically smaller and less advanced than military drones. Examples of scientific drones include IceBridge, TDR-SS, and Global Hawk.

Civilian drones, also known as consumer drones or hobby drones, are used for a variety of purposes, including photography, videography, and delivery services. They are typically smaller and less advanced than military and scientific drones. Examples of civilian drones include the DJI Phantom, DJI Mavic, Parrot Bebop, Yuneec Typhoon, and Autel Robotics EVO.

2

Sec: 1 Military Drones

Military drones are unmanned aerial vehicles (UAVs) designed for use by military and security forces for reconnaissance, surveillance, and offensive operations. They are equipped with a variety of sensors, cameras, communication systems, and other payloads to support their mission requirements. Some military drones are armed with missiles, bombs, or other weapons to support offensive operations.

3

Ch: 1 MQ-9 Reaper (Predator B)

Known specs

- **Unit Cost:** $56.5 million (includes four aircraft with sensors, ground control station and Predator Primary satellite link) (fiscal 2011 dollars)
- **Thrust:** 900 shaft horsepower maximum
- **Wingspan:** 66 feet (20 meters)
- **Length:** 36 feet (11 meters)
- **Height:** 12.5 feet (3.8 meters)
- **Maximum takeoff weight:** 14,000 pounds (6,350 kg)
- **Maximum speed:** 240 knots (277 mph, 446 km/h)
- **Range:** 1,150 miles (1,000 nautical miles)| ER:1,611 miles (1,400 nautical miles)
- **Fuel Capacity:** 4,000 pounds (602 gallons) | ER: 6,000 pounds (903 gallons)
- **Ceiling:** Up to 50,000 feet (15,240 meters)
- **Endurance:** 27 hours
- **Crew (remote):** Pilot & Sensor Operator
- **Payload capacity:** 3,750 pounds (1,700 kg)
- **Armament:**

1. Combination of AGM-114 Hellfire missiles
2. GBU-12 Paveway II
3. GBU-38 Joint Direct Attack Munitions
4. GBU-49 Enhanced Paveway II
5. GBU-54 Laser Joint Direct Attack Munitions

History and usage

The MQ-9 Reaper, also known as the Predator B, is a military drone developed by General Atomics Aeronautical Systems. It was first used in combat by the United States Air Force in 2007. The MQ-9 Reaper is primarily used for reconnaissance, surveillance, and targeted killings. It has been used in a variety of conflicts, including the War in Afghanistan and the War in Pakistan. The USAF operated over 300 MQ-9 Reapers as of May 2021,[8] with 16 additional units on the way as authorized by the FY2021 Congressional budget.

Additional information

- The MQ-9 Reaper is often armed with Hellfire missiles and laser-guided bombs.
- It can be controlled remotely or fly autonomously using a pre-programmed flight plan.
- The MQ-9 Reaper has a longer endurance and payload capacity than its predecessor, the MQ-1 Predator.
- The U.S. Air Force has the largest fleet of MQ-9 Reaper drones, with over 60 in active service.
- Other countries, such as the United Kingdom, France, and Italy, also operate the MQ-9 Reaper in their militaries.

9

4

Ch: 2 MQ-1B Predator

Known specs

- **Unit cost:** $20 million (includes four aircraft with sensors, ground control station and Predator Primary satellite link) (fiscal 2009 dollars)
- **Thrust:** 115 Horsepower
- **Wingspan:** 55 feet (16.8 meters)
- **Length:** 27 feet (8.2 meters)
- **Height:** 7 feet (2.1 meters)
- **Maximum takeoff weight:** 2,250 pounds (1,021 kg)
- **Maximum speed:** 84 knots (97 mph, 156 km/h)
- **Range:** 770 miles (675 nautical miles)
- **Fuel Capacity:** 665 pounds (100 gallons)
- **Ceiling:** 25,000 feet (7,620 meters)
- **Endurance:** 24 hours
- **Crew (remote):** Pilot & Sensor Operator
- **Payload capacity:** 450 pounds (204 kg)
- **Armament:** Two laser-guided AGM-114 Hellfire missiles

History and usage

The MQ-1 Predator is a military drone developed by General Atomics Aeronautical Systems. It was first used in combat by the United States Air Force in 1995. The Predator is primarily used for reconnaissance and surveillance missions. It has been used in a variety of conflicts, including the War in Afghanistan and the Iraq War.

Additional information

- The Predator is armed with AGM-114 Hellfire missiles and is capable of carrying 2 missiles at a time.
- It can be controlled remotely or fly autonomously using a pre-programmed flight plan.
- The Predator is considered to be the first armed drone to be used in combat by the United States.
- The U.S. Air Force has retired the Predator in favor of the MQ-9 Reaper.
- Other countries, such as Italy, also operate the Predator in their militaries.
- The Predator drone is the first unmanned aircraft to be certified by the Federal Aviation Administration for routine commercial use in the United States.

5

CH: 3 RQ-4 Global Hawk

Known specs

- **Unit Cost:** $220–250 Million USD (includes the cost of the airframe, the sensors and other payloads, as well as the ground control equipment and support services)
- **Thrust:** 7,600 pounds
- **Wingspan:** 130.9 feet (39.9 meters)
- **Length:** 47.6 feet (14.5 meters)
- **Height:** 15.4 feet (4.7 meters)
- **Maximum takeoff weight:** 32,250 pounds (14,600 kg)
- **Maximum speed:** 357 mph (305 knots, 565 km/h)
- **Range:** 12,300 nautical miles
- **Fuel Capacity:** 17,300 pounds (7847 kilograms)
- **Ceiling:** 60,000 feet (18,288 meters)
- **Endurance:** 34 hours
- **Crew (remote):** Three (LRE pilot, MCE pilot, and sensor operator)
- **Payload capacity:** 3,000 pounds (1,360 kg)
- **Armament:** None

History and usage

The RQ-4 Global Hawk is a high-altitude, long-endurance military drone developed by Northrop Grumman. It was first used by the United States Air Force in 2001. The RQ-4 Global Hawk is primarily used for reconnaissance and surveillance missions. It can fly at altitudes of up to 60,000 feet (18,288 meters) and has a range of around 12,300 nautical miles (22,222 km). It is equipped with a variety of sensors including synthetic aperture radar, electro-optical/infrared camera and signals intelligence systems.

Additional information

- The RQ-4 Global Hawk has been used in a variety of conflicts, including the War in Afghanistan and the Iraq War.
- The U.S. Air Force has the largest fleet of RQ-4 Global Hawk drones, with over 30 in active service.
- The RQ-4 Global Hawk has been used for civilian purposes such as monitoring natural disasters, mapping and border surveillance.
- The RQ-4 Global Hawk was originally developed as a replacement for the U-2 spy plane.
- Other countries such as Australia and South Korea are also operate the RQ-4 Global Hawk in their militaries.

6

Ch: 4 RQ-21 Blackjack

Known specs

- **Unit Cost:** $10–15 Million USD
- **Thrust:** Not publicly available
- **Wingspan:** 15 feet (4.6 meters)
- **Length:** 8.2 feet (2.5 meters)
- **Height:** 3.3 feet (1 meter)
- **Maximum takeoff weight:** 130 pounds (59 kg)
- **Maximum speed:** 87 knots - 100 mph (170 km/h, 90 kn)
- **Range:** 58 mi (93 km, 50 nmi)
- **Fuel Capacity:** Not publicly available
- **Ceiling:** 19,500 ft (5,900 m)
- **Endurance:** 16 hours
- **Crew (remote):** Not publicly available
- **Payload capacity:** 30 pounds (14 kg)
- **Armament:** None

History and usage

The RQ-21 Blackjack is a small tactical unmanned aircraft system (STUAS) developed by Insitu, a subsidiary of Boeing. It was first used by the United States Navy in 2012. The Blackjack is primarily used for reconnaissance and surveillance missions. It has been used in a variety of operations, including counter-piracy and counter-narcotics.

Additional information

- The Blackjack is launched by a pneumatic launcher and recovered by a "Skyhook" system.
- It can be controlled remotely or fly autonomously using a pre-programmed flight plan.
- The Blackjack has a small footprint and is designed to be launched and recovered from ships and small coastal bases.
- The U.S. Navy has procured more than 100 RQ-21 Blackjack systems.
- The RQ-21 Blackjack has been used by the U.S. Marine Corp and U.S. Special Operations Command.
- It is also used for civilian purposes such as border surveillance, search and rescue, and mapping.

7

CH: 5 RQ-7 Shadow

Photo by: Sgt. Eric Warren

Known specs

- **Unit Cost:** $1-2 Million USD
- **Thrust:** Not publicly available
- **Wingspan:** 10 feet (3.05 meters)
- **Length:** 15 feet (4.57 meters)
- **Height:** 5 feet (1.52 meters)
- **Maximum takeoff weight:** 130 pounds (59 kilograms)
- **Maximum speed:** 87 Knots - 100 mph (160 km/h)
- **Range:** 150 miles (240 kilometers)
- **Fuel Capacity:** Not publicly available
- **Ceiling:** 14,000 feet (4,267 meters)
- **Endurance:** 9 Hours
- **Crew (remote):** Not publicly available
- **Payload capacity:** 40 pounds (18 kilograms)
- **Armament:** None

History and usage

The RQ-7 Shadow has been used by various military and security forces for reconnaissance, surveillance, and target acquisition missions. It has been deployed in various theaters of operation, including Iraq and Afghanistan.

Additional information

The RQ-7 Shadow is a small, tactical unmanned aerial vehicle (UAV) developed by AAI Corporation, now part of Textron Systems. It is designed to provide real-time intelligence, surveillance, and reconnaissance (ISR) capabilities for ground troops in support of forward operations. The RQ-7 Shadow has undergone several upgrades and modifications to improve its capabilities and performance, and it remains a widely used UAV in military and security applications.

8

Sec: 2 Scientific Drones

Scientific drones are unmanned aerial vehicles designed for use in scientific research and exploration. They are equipped with specialized sensors and instruments to collect data on

various environmental and scientific phenomena, such as atmospheric chemistry, wildlife migration patterns, and geological
structures. Scientific drones are typically operated by research
organizations, universities, and government agencies.

9

Ch: 1 Nasa Global Hawk

Known specs

- **Unit Cost:** $120 million to $150 million USD
- **Thrust:** 11,000 pounds of thrust
- **Wingspan:** 116.2 feet (35.4 meters)
- **Length:** 47.6 feet (14.5 meters)
- **Height:** 15.4 feet (4.7 meters)
- **Maximum takeoff weight:** 26,250 pounds (11,880 kilograms)
- **Maximum speed:** 460 knots (529 miles per hour or 850 kilometers per hour)
- **Range:** 11,000 nautical miles (12,700 miles or 20,400 kilometers)
- **Fuel Capacity:** 3,200 gallons (12,100 liters)
- **Ceiling:** 60,000 feet (18 kilometers)
- **Endurance:** 32 Hours
- **Crew (remote):** Not publicly available
- **Payload capacity:** Not publicly available
- **Armament:** None

History and usage

The NASA Global Hawk is a high-altitude, long-endurance unmanned aircraft developed by NASA, the National Aeronautics and Space Administration of the United States. The Global Hawk is used for scientific research and data collection, such as atmospheric monitoring and mapping. It can fly at an altitude of 60,000 feet for up to 28 hours, providing a cost-effective and efficient solution for scientific research.

Additional information

- The NASA Global Hawk is equipped with a range of sensors, including cameras, spectrometers, and atmospheric sensors, to collect data for scientific research.
- The NASA Global Hawk has been used for a variety of research projects, including monitoring and mapping the atmosphere, studying weather patterns, and tracking wildlife.
- NASA Global Hawk have been used for various missions like studying hurricanes, typhoons and other natural disasters, studying the ozone layer and other scientific research.

10

Ch: 2 SENTRY

Known specs

- **Unit Cost:** Not publicly available
- **Thrust:** Not publicly available

- **Dimensions:** 20" x 20" x 10.6" (50 cm x 50 cm x 27 cm)
- **Maximum takeoff weight:** 6.8 lbs
- **Maximum speed:** Not publicly available
- **Range:** 200 ft (60m) tether
- **Fuel Capacity:** Supplied by tether with backup battery
- **Ceiling:** 200 ft (60m) tether
- **Endurance:** Tethered
- **Crew (remote):** 1
- **Payload capacity:** 1.8 lbs
- **Armament:** None

History and usage

The Hoverfly SENTRY drone is a multirotor UAV developed by Hoverfly Technologies for industrial inspection and monitoring missions. It is equipped with high-resolution cameras and other sensors to capture detailed images and data of structures and assets being inspected.

Additional information

The SENTRY drone is designed for ease of use and versatility, with a compact and lightweight design well-suited for use in a range of environments and conditions. It provides businesses with a cost-effective and efficient way to gather data and information about their critical assets and infrastructure.

11

Ch: 3 AeroVironment Global Observer

Known specs

- **Unit Cost:** Not publicly available
- **Thrust:** Not publicly available
- **Wingspan:** 175 feet (53 m)
- **Length:** 70 feet (21 m)
- **Height:** Not publicly available
- **Maximum takeoff weight:** Not publicly available
- **Maximum speed:** Not publicly available
- **Range:** Global

- **Fuel Capacity:** Liquid hydrogen-powered internal combustion power plant
- **Ceiling:** 65,000 feet (17,000–20,000 m)
- **Endurance:** 5–7 days
- **Crew (remote):** Not publicly available
- **Payload capacity:** Up to 400 lbs (180 kg)
- **Armament:** None

History and usage

The Global Observer is a high-altitude, long-endurance unmanned aircraft developed by AeroVironment, an American company. The Global Observer was designed for scientific research and data collection, such as atmospheric monitoring and mapping. It can fly at an altitude of 65,000 feet for up to 5 days, providing a cost-effective and efficient solution for scientific research.

Additional information

- The Global Observer is powered by hydrogen fuel cells, which allows for a longer flight time than traditional gasoline-powered drones.
- The Global Observer is equipped with a range of sensors, including cameras, spectrometers, and atmospheric sensors, to collect data for scientific research.
- The Global Observer has been used for a variety of research

projects, including monitoring and mapping the atmosphere, studying weather patterns, and tracking wildlife.
· The first flight of the Global Observer using hydrogen fuel
occurred on 11 January 2011.

12

CH: 4 Boreal Lab UAV

Known specs

- **Unit Cost:** Not publicly available
- **Thrust:** Not publicly available
- **Wingspan:** 4.20 Meters
- **Length:** Not publicly available

- **Height:** Not publicly available
- **Maximum takeoff weight:** 55lbs
- **Maximum speed:** 70 knots – 130 kmh
- **Range:** 700km
- **Fuel Capacity:** 30 Litre
- **Ceiling:** 3500m
- **Endurance:** 8 hours
- **Crew (remote):** 2 Operators
- **Payload capacity:** 5Kg
- **Armament:** None

History and usage

The Boreal LabUAV is a drone developed by the French company Airbus Defence and Space, and is designed for a range of scientific and research applications. It has been used in various missions, including monitoring of natural disasters, atmospheric research, and environmental monitoring. The drone is known for its ability to operate in challenging environments and provide high-quality data and imaging for scientific analysis.

Additional information

The Boreal LabUAV is a highly versatile drone that can be configured for a wide range of scientific and research missions. It is capable of operating in remote and challenging environments, including Arctic and Antarctic regions, and provides high-quality data and images for scientific analysis. The drone is equipped with a range of sensors and instruments that can be customized to meet specific mission requirements, including cameras, spectrometers, laser altimeters, and radiometers.

In addition to its scientific capabilities, the Boreal LabUAV is also designed for ease of use and maintenance, with a modular design that allows for quick and easy replacement of components in the field. This makes it a cost-effective solution for a range of scientific research missions, and has helped establish it as a leader in the field of atmospheric research and environmental monitoring.

Overall, the Boreal LabUAV is a reliable and versatile drone that provides valuable data and images for scientific analysis and research, and has proven to be a valuable tool for understanding the earth's environment and natural processes.

13

Ch: 5 Aeryon Scout

Known specs

- **Unit Cost:** No longer for Sale
- **Thrust:** Not publicly available
- **Wingspan:** Not publicly available
- **Length:** 0.8 m (2 ft 7 in)
- **Height:** 0.30 m (1 ft 0 in)
- **Maximum takeoff weight:** 1.7 kg (3.74 lb)
- **Maximum speed:** 50 kmh
- **Range:** 3 km (1.9miles)
- **Fuel Capacity:** Not publicly available
- **Ceiling:** 333 m (1,000 ft) AGL 5,000 m (15,000 ft) ASL
- **Endurance:** 24 minutes
- **Crew (remote):** 1 Pilot
- **Payload capacity:** Approximately 1 kg (2.2 lbs)
- **Armament:** None

History and usage

The Aeryon Scout Drone was developed by Aeryon Labs, a Canadian company, as a small, lightweight, and highly maneuverable drone for a range of environmental monitoring and research missions. It is equipped with high-resolution cameras and other sensors that allow it to gather detailed images and data for a wide range of applications, including wildlife surveys, oil spill monitoring, and natural disaster response.

Additional information

The Aeryon Scout Drone is designed for ease of use, with a simple and intuitive remote control system that allows it to be quickly and easily operated by a single pilot. It is also designed for durability and reliability, with a rugged and compact design that makes it well suited for use in challenging environments and remote locations.

Overall, the Aeryon Scout Drone is a highly versatile and capable drone that has proven to be a valuable tool for environmental monitoring and research, and has been used by a range of organizations, including government agencies, universities, and non-profit organizations.

14

Sec: 3 Civilian Drones

Civilian drones are unmanned aerial vehicles designed for personal or commercial use. They are equipped with cameras, sensors, and other payloads to support various applications, such as aerial photography and videography, real estate photography, agriculture, and search and rescue operations. Civilian drones are typically operated by individuals, businesses, or public safety organizations.

Ch: 1 Phantom 4 Pro V2.0

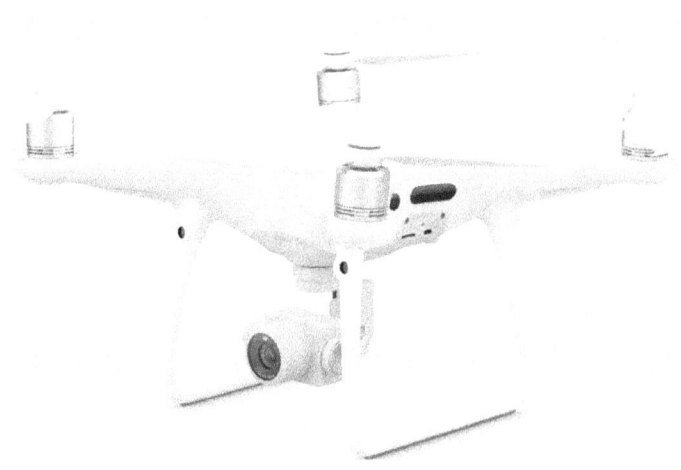

Known specs

- **Unit Cost:** $1,599 USD
- **Thrust:** Not publicly available

- **Diagonal Size (Propellers Excluded):** 350 mm
- **Maximum takeoff weight:** Not publicly available
- **Maximum speed:**S-mode: 45 mph (72 kph)
- **Range:** 6000m
- **Fuel Capacity:** Battery powered
- **Ceiling:** 19685 ft (6000 m)
- **Endurance:** 30 minutes
- **Crew (remote):** 1
- **Payload capacity:** Not publicly available
- **Armament:** None

History and usage

The DJI Phantom is a series of quadcopter drones developed and manufactured by DJI, a Chinese technology company. The first Phantom drone was released in January 2013, and since then, several versions have been released, including the Phantom 2, Phantom 3, Phantom 4, and Phantom 4 Pro. The Phantom is widely used by hobbyists and professionals alike for aerial photography and videography, surveying, and mapping. It is known for its ease of use, portability, and advanced features such as GPS and obstacle avoidance.

Additional information

- The Phantom series is one of the most popular consumer drones in the world, with millions sold worldwide.
- DJI offers a range of accessories for the Phantom series, including different camera options, and gimbal stabilizers for better image stabilization.

- The Phantom series can be controlled using a remote controller or a mobile device such as a smartphone or tablet.
- Some versions of the Phantom series include advanced features such as a 3-axis gimbal, 4K camera, and the ability to fly autonomously using waypoints.
- DJI also offers a Phantom 4 Pro version with a thermal imaging camera and software for use in search and rescue missions, inspections, and surveying of powerlines and pipelines.
- The Phantom series has been used by various industries such as real estate, agriculture, and film production.

16

Ch: 2 DJI Mavic 3

Known specs

- **Unit Cost:** $2,049 USD
- **Thrust:** Not publicly available

- **Unfolded (without propellers):** 347.5×283×107.7 mm
- **Diagonal Length:** 380.1 mm
- **Maximum takeoff weight:** Not publicly available
- **Maximum speed:** 21 m/s (S mode)
- **Range:** 30 km
- **Fuel Capacity:** Battery Powered
- **Ceiling:** 6000 m
- **Endurance:** 40-46 Minutes
- **Crew (remote):** 1
- **Payload capacity:** Not publicly available
- **Armament:** None
-
- Wingspan: 16.6 inches (420mm)
- Length: 7.8 inches (198mm)
- Height: 3.3 inches (83mm)
- Maximum takeoff weight: 1.64 lbs (743g)
- Maximum speed: 40mph (64km/h)
- Flight time: Up to 27 minutes
- Camera: 4K video and 12MP still photos

History and usage

The DJI Mavic is a series of compact and foldable drones developed and manufactured by DJI. The Mavic Pro was first released in September 2016, followed by the Mavic Air in January 2018 and the Mavic 2 in August 2018. The Mavic series is known for its portability and advanced features such as 4K camera, 3-axis gimbal, and obstacle avoidance. It is widely used by hobbyists and professionals for aerial photography and videography, as well as for surveying and mapping.

Additional information

- The Mavic series is known for its compact and portable design, making it easy to take on the go.
- The Mavic Pro and Mavic Air can be controlled using a remote controller or a mobile device such as a smartphone or tablet.
- The Mavic 2 includes advanced features such as a 4K camera, 3-axis gimbal, and obstacle avoidance.
- DJI offers a range of accessories for the Mavic series, including different camera options, gimbal stabilizers, and prop guards.
- The Mavic series has been used by various industries such as real estate, agriculture, and film production.
- The Mavic series is also used for search and rescue missions, inspections, and surveying of powerlines and pipelines.

17

Ch: 3 Parrot Bebop

Known specs

- **Unit Cost:** $150-200 USD
- **Thrust:** Not publicly available
- **Wingspan:** 9.76 in
- **Length:** 14.96
- **Height:** 1.42 in
- **Maximum takeoff weight:**
- **Maximum speed:** 13 m/s

- **Range:** 1.4 miles
- **Fuel Capacity:** Battery powered
- **Ceiling:** 10m
- **Endurance:** 11 minutes
- **Crew (remote):** 1
- **Payload capacity:** Not publicly available
- **Armament:** none

History and usage

The Parrot Bebop is a series of compact and lightweight drones developed and manufactured by Parrot. The Bebop was first released in December 2013, followed by the Bebop 2 in December 2014. The Bebop series is known for its compact size, high-quality camera, and easy-to-use controls. It is widely used by hobbyists and professionals for aerial photography and videography, as well as for surveying and mapping.

Additional information

- The Bebop series is known for its compact and lightweight design, making it easy to take on the go.
- The Bebop can be controlled using a remote controller or a mobile device such as a smartphone or tablet.
- The Bebop 2 includes advanced features such as a 14MP camera, 3-axis digital stabilization, and GPS.
- Parrot offers a range of accessories for the Bebop series, including different camera options, gimbal stabilizers, and

prop guards.
- The Bebop series has been used by various industries such as real estate, agriculture, and film production.
- The Bebop series is also used for search and rescue missions, inspections, and surveying of powerlines and pipelines.

18

Ch: 4 Yuneec Typhoon H Plus

Known specs

- **Unit Cost:** $1,999
- **Thrust:** Not publicly available
- **Dimensions:** 115 x 80 x 130 mm
- **Maximum takeoff weight:** Not publicly available
- **Maximum speed:** 30 mph
- **Range:** 1 mile
- **Fuel Capacity:** Battery powered
- **Ceiling:** 1640 ft
- **Endurance:** Up to 28 min (with C23)
- **Crew (remote):** 1 or 2
- **Payload capacity:** Not publicly available
- **Armament:** None

History and usage

The Yuneec Typhoon is a series of drones developed and manufactured by Yuneec International, a Chinese company. The first Typhoon model was released in 2013, followed by several updates and new models. The Typhoon series is known for its stability, long flight time, and high-quality camera. It is widely used by hobbyists and professionals for aerial photography and videography, as well as for surveying and mapping.

Additional information

- The Typhoon series offers a range of models with different camera options and flight modes.
- The Typhoon can be controlled using a remote controller or a mobile device such as a smartphone or tablet.
- Some models in the Typhoon series include obstacle avoidance technology, making it easier to fly in tight spaces.
- Yuneec offers a range of accessories for the Typhoon series, including different camera options, gimbal stabilizers, and prop guards.
- The Typhoon series is also used for search and rescue missions, inspections, and surveying of powerlines and pipelines.
- The Typhoon series is a popular choice among hobbyists and professionals, due to its stability, long flight time, and high-quality camera.

19

Ch: 5 Autel Robotics EVO II Pro V3

Known specs

- **Unit Cost:** $2,500 USD
- **Thrust:** Not publicly available
- **Dimensions:** 9.1*5.1*4.3 inches (folded); 18*22*4.3 inches (unfolded)
- **Maximum takeoff weight:** 4.41 lbs (1999 g⊠
- **Maximum speed:** 27 mph
- **Range:** 5 miles
- **Fuel Capacity:** Battery powered
- **Ceiling:** 7000 m
- **Endurance:** 35–40 minutes
- **Crew (remote):** 1
- **Payload capacity:** Not publicly available
- **Armament:** None

History and usage

The Autel Robotics EVO is a consumer drone developed and manufactured by Autel Robotics, an American company. The EVO was first released in 2018, and it's known for its portability, endurance and long flight time. It's a popular choice among hobbyists and professionals for aerial photography and videography, as well as for surveying and mapping.

Additional information

- The EVO offers a range of features such as obstacle avoidance technology, which makes it easier to fly in tight spaces.
- The EVO can be controlled using a remote controller or a mobile device such as a smartphone or tablet.
- Autel Robotics offers a range of accessories for the EVO, including different camera options, gimbal stabilizers, and prop guards.
- The EVO is also used for search and rescue missions, inspections, and surveying of powerlines and pipelines.
- The EVO is a popular choice among hobbyists and professionals, due to its portability, endurance and long flight time.
- The EVO has a 3-axis gimbal and 6K Camera, which allows for smooth and stable video footage.
- The EVO has 360° obstacle avoidance with 19 groups of sensors.

About the Author

Avery Wright brings a unique and important perspective to the world of AI. With a focus on the ethical, legal, and social implications, Avery provides valuable insights and inspiration to those interested in this rapidly-evolving technology.

You can connect with me on:
🌐 https://sirexodia.wixsite.com/avery-wright
🐦 https://twitter.com/AveryWrightAI
📘 https://www.facebook.com/profile.php?id=100089987171726

Subscribe to my newsletter:
✉ https://sirexodia.wixsite.com/avery-wright

Also by Avery Wright

"From Predator to Phantom: A History of Drones" is a comprehensive guide to the evolution of drones. The book covers military, scientific, and civilian drones, each chapter focusing on a different drone with detailed specs, history, usage, and additional info. Get a complete look at the impact drones have had on society and the advancements made in drone technology. A must-read for drone enthusiasts and technology buffs.

Mastering Midjourney AI - The Beginner's Handbook
Mastering Midjourney AI: The Beginner's Handbook is a comprehensive guide for beginners looking to learn about the Midjourney AI platform and how to use it for image generation. The book covers a range of topics, including understanding Midjourney AI's parameters and settings, using URLs for image inspiration, adjusting image quality, and more.
https://www.amazon.com/dp/B0BV8PGDXT

Transformative Art - A Journey with Artificial Intelligence

Transformative Art: A Journey with AI is a visually stunning and thought-provoking book that explores the intersection of artificial intelligence and the world of art. The book features breathtaking images of futuristic cities, technology, vehicles, robots, flying ships, conceptual art, abstract art, and unique pieces, all within the context of transformative art. Each chapter begins with a powerful quote that sets the tone for a deep dive into the themes of perception, change, reflection, risk-taking, emotional connection, the journey within, and the universal language of art. The book is written by Avery Wright, a talented author with a passion for exploring the ways in which technology is changing our lives and our world. This book is a must-read for anyone interested in the intersection of art, technology, and the human experience. https://www.amazon.com/dp/B0BTWNYLJD